SPACE FRONTIERS

Space Missions

Helen Whittaker

A+

This edition first published in 2011 in the United States of America by Smart Apple Media.

Smart Apple Media
P.O. Box 3263
Mankato, MN, 56002

First published in 2010 by
MACMILLAN EDUCATION AUSTRALIA PTY LTD
15–19 Claremont Street, South Yarra 3141

Visit our website at www.macmillan.com.au or go directly to www.macmillanlibrary.com.au

Associated companies and representatives throughout the world.

Copyright © Macmillan Publishers Australia 2010

National Library of Australia Cataloguing-in-Publication entry

Library of Congress Cataloging-in-Publication Data

Whittaker, Helen, 1965-
    Space missions / Helen Whittaker.
        p. cm. — (Space frontiers)
    Includes index.
    ISBN 978-1-59920-573-1 (lib. bdg.)
    1. Outer space—Exploration—Juvenile literature. I. Title.
    QB500.262.W49 2011
919.904—dc22

                            2009038479

Edited by Laura Jeanne Gobal
Text and cover design by Cristina Neri, Canary Graphic Design
Page layout by Cristina Neri, Canary Graphic Design
Photo research by Brendan and Debbie Gallagher
Illustrations by Alan Laver

Manufactured in China by Macmillan Production (Asia) Ltd.
Kwun Tong, Kowloon, Hong Kong
Supplier Code: CP December 2009

Acknowledgments
The author and the publisher are grateful to the following for permission to reproduce copyright material:

Front cover photos of space shuttle *Atlantis* lift-off courtesy of NASA/Kennedy Space Center; blue nebula background © sololos/iStockphoto.

Photographs courtesy of:
Jean-Leon Huens/*National Geographic*/Getty Images, **19** (bottom); Michael Rougier/Time Life Pictures/Getty Images, **8** (bottom); Calvin J. Hamilton, **20**; NASA, **12**, **21**, back cover; NASA, ESA, and The Hubble Heritage Team (STScI), **19** (top); NASA/Goddard Space Flight Center Scientific Visualization Studio, **22**; NASA/HSF, **5**, **13**, **16**, **17**, **30**; NASA/Johns Hopkins University Applied Physics Laboratory/Carnegie Institution of Washington, **25**; NASA/ JPL, **23**; NASA/JPL-Caltech/Potsdam Univ, **6–7**; NASA/JSC, **11**; NASA/JSC, Eugene Cernan, **4**; NASA/Kennedy Space Center, **3**; NASA photo by Lori Losey, **15**; NASA/MSFC, **8** (top), **9**, **18**; Photolibrary/SPL, **27** (bottom); Photolibrary/ NASA/SPL, **26**, **27** (top); Photolibrary/RIA NOVOSTI/SPL, **10**, **24**; SOHO (ESA & NASA), **28**.

Images used in design and background on each page © prokhorov/iStockphoto, Soubrette/iStockphoto.

CONTENTS

Glossary Words

When a word is printed in **bold**, you can look up its meaning in the Glossary on page 31.

SPACE FRONTIERS

A frontier is an area that is only just starting to be discovered. Humans have now explored almost the entire planet, so there are very few frontiers left on Earth. However, there is another frontier for us to explore and it is bigger than we can possibly imagine—space.

Where Is Space?

Space begins where Earth's **atmosphere** ends. The atmosphere thins out gradually, so there is no clear boundary marking where space begins. However, most scientists define space as beginning at an altitude of 62 miles (100 km). Space extends to the very edge of the universe. Scientists do not know where the universe ends, so no one knows how big space is.

Exploring Space

Humans began exploring space just by looking at the night sky. The invention of the telescope in the 1600s and improvements in its design have allowed us to see more of the universe. Since the 1950s, there has been another way to explore space—spaceflight. Through spaceflight, humans have **orbited** Earth, visited the Moon, and sent space probes, or small unmanned spacecraft, to explore our **solar system**.

Spaceflight is one way of exploring the frontier of space. Astronaut Harrison Schmitt collects Moon rocks during the Apollo 17 mission in December 1972.

SPACE MISSIONS

A space mission involves sending machines into space in order to achieve one or more objectives. It can also involve sending people into space.

Mission Objectives

Space missions have some common objectives:

- exploring the solar system
- launching and repairing space telescopes and communications **satellites** (for telephone, television, radio, and Internet use)
- launching space probes such as **landers** and **rovers**
- constructing space stations
- conducting scientific research
- developing military technology such as spy satellites and antimissile systems
- preparing for future space missions

In the future, mission objectives may also include the following:

- developing space tourism
- setting up a base on the Moon
- mining planets and moons for their natural resources
- sending humans to live on other planets

On space shuttle mission STS-112 in 2002, American astronaut David A. Wolf attaches a camera to the exterior of the International Space Station.

Mission Fact!

Space exploration is a controversial subject, mainly because it is very expensive. Some people argue that the costs outweigh the benefits, and that money and resources spent on space missions would be better used elsewhere.

A TIMELINE OF SPACE MISSIONS

Since the launch of the first artificial **satellite**, Sputnik 1, in 1957, thousands of spacecraft have completed successful missions in space. This timeline shows some of the most interesting and noteworthy space missions. For more information about each mission, refer to the page numbers in brackets.

	1950	1960	1970

Earth Orbit

Sputnik 1
1957
[see page 8]

Vostock 1
1961
[see page 9]

Project Mercury
1959-63
[see page 8]

Project Gemini
1965-66
[see pages 10–11]

The Moon and Mars

Luna Program (the Moon) 1959–76
[see page 10]

Apollo Program (the Moon)
1961–75 [see pages 12–13]

Mariner Program (Mars, Venus, and Mercury)
1962–75 (see page 20–21)

Other Planets

Venera Probes (Venus) 1961–84
[see page 24]

Other Destinations

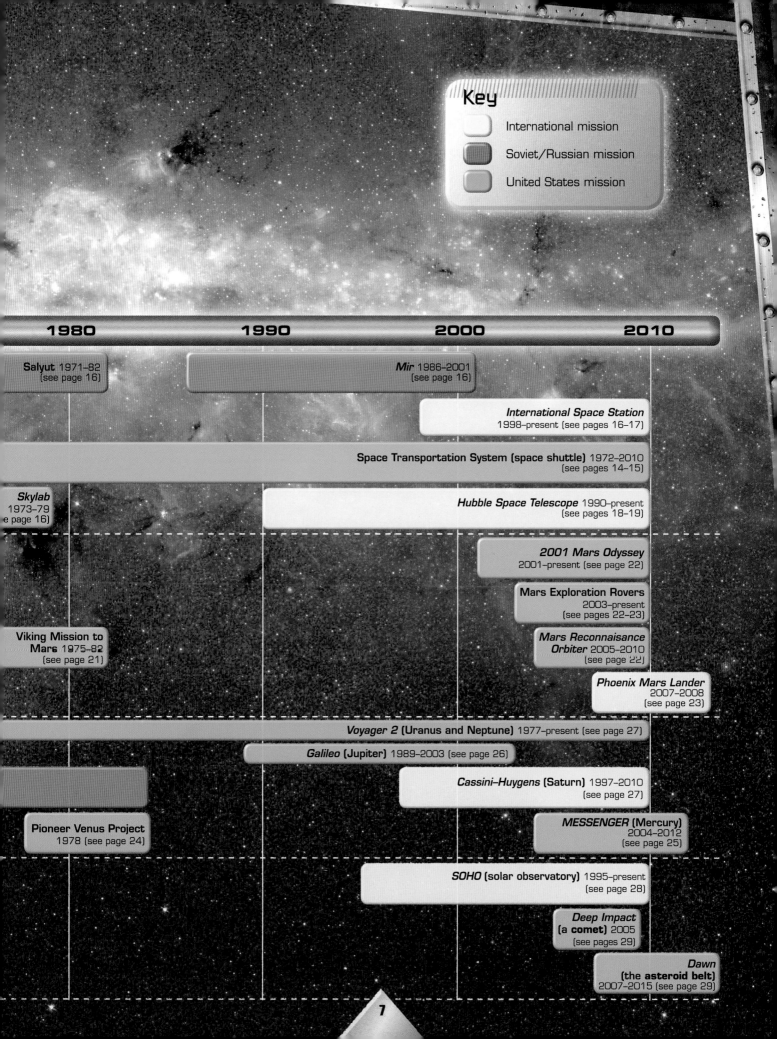

Key

☐ International mission

▨ Soviet/Russian mission

▨ United States mission

1980   **1990**   **2000**   **2010**

Salyut 1971–82
(see page 16)

Mir 1986–2001
(see page 16)

International Space Station
1998–present (see pages 16–17)

Space Transportation System (space shuttle) 1972–2010
(see pages 14–15)

Skylab
1973–79
e page 16)

Hubble Space Telescope 1990–present
(see pages 18–19)

2001 Mars Odyssey
2001–present (see page 22)

Mars Exploration Rovers
2003–present
(see pages 22–23)

*Mars Reconnaisance
Orbiter* 2005–2010
(see page 22)

Phoenix Mars Lander
2007–2008
(see page 23)

**Viking Mission to
Mars** 1975–82
(see page 21)

Voyager 2 (Uranus and Neptune) 1977–present (see page 27)

Galileo (Jupiter) 1989–2003 (see page 26)

Cassini–Huygens (Saturn) 1997–2010
(see page 27)

MESSENGER (Mercury)
2004–2012
(see page 25)

Pioneer Venus Project
1978 (see page 24)

SOHO (solar observatory) 1995–present
(see page 28)

Deep Impact
(a comet) 2005
(see pages 29)

Dawn
(the asteroid belt)
2007–2015 (see page 29)

THE RACE INTO SPACE

MISSION OBJECTIVE: To send machines and humans into space

Following World War II (1939–45), the United States and the Soviet Union became fierce rivals who aimed to be better prepared for the next war by building bigger and better weapons.

The Space Race

Developments in missile technology led to rockets that could travel much farther, as far as space. Both countries began to encourage research into space technology for three main reasons:

- Satellites in space could spy on other countries.
- The technology required for space travel could be used in weapons on Earth.
- Accomplishments in space, the final frontier, would boost the morale of citizens still recovering from years of war.

This led to the **Space Race**.

Alan Shepard (1923–98)

Mercury astronaut Alan Shepard was the first American in space, flying *Freedom 7* on May 5, 1961. He was grounded for most of the 1960s due to an inner ear complaint. In 1971, he commanded the Apollo 14 mission and became the fifth person to walk on the Moon.

Sputnik 1

On October 4, 1957, the Soviet Union announced the successful launch of *Sputnik 1*, the first artificial satellite. The news came as a shock to the rest of the world, especially the United States, which had thought it was ahead in the Space Race. In 1958, responding to *Sputnik 1*, the American government formed the National Aeronautics and Space Administration (NASA).

Project Mercury

The next goal in the Space Race was to send the first human into space. In 1959, NASA launched Project Mercury, the United States's first manned spaceflight program. Just one month before the first manned Mercury flight was due to take place, news came that the Soviet Union had got there first.

A replica of *Sputnik 1* is displayed at the 1958 Brussels World's Fair in Belgium. When in space, the satellite's antennae broadcast a continuous beeping sound that proved the mission's success to the rest of the world.

Vostok 1

On April 12, 1961, Vostok 1 astronaut Yuri Gagarin orbited Earth once in the spacecraft *Swallow* and landed safely. The flight lasted less than two hours. Gagarin was awarded dozens of medals from countries around the world. The 12th of April is still a public holiday in Russia.

▼ **Vostok 1's success was front-page news all over the world. *The Huntsville Times* is from Alabama.**

Mission Fact!

For a manned spaceflight to be officially recognised in 1961, the astronaut had to land with the spacecraft. According to some sources, Yuri Gagarin ejected from his spacecraft prior to landing and parachuted back to Earth. The Soviet Union, however, claimed Gagarin landed with the spacecraft.

The Huntsville Times

VOL. 31, NO. 21   CHICAGO DAILY NEWS SERVICE   HUNTSVILLE, ALABAMA, WEDNESDAY, APR. 12, 1961   ASSOCIATED PRESS — WIREPHOTO   45c PER WEEK

23 PAGES TODAY   Where Progress... Covers The Valley!

Man Enters Space

'So Close, Yet So Far,' Sighs Cape

U. S. Had Hoped For Own Launch

Hobbs Admits 1944 Slaying

This is Russian Maj. Yuri Gagarin, history's first man in space. The Russians today rocketed him around the earth in one orbit taking slightly less than 90 minutes and brought him back safely to a prearranged spot in the Soviet Union. (AP Wirephoto via radio from Moscow)

Praise Is Heaped On Major Gagarin

'Worker' Stands By Story

First Man To Enter Space Is 27, Married, Father Of Two

Soviet Officer Orbits Globe In 5-Ton Ship

Maximum Height Reached Reported As 188 Miles

VON BRAUN'S REACTION:

To Keep Up, U.S.A. Must Run Like Hell'

WERNHER VON BRAUN
He Praises A Russian Achievement
By BILL AUSTIN
Of The Times Staff

Reds Deny Spacemen Have Died

By THE ASSOCIATED PRESS

No Astronaut Signal Received At Ft. Monmouth

Reds Win Running Lead In Rac   To Control Space

9

DESTINATION: THE MOON

Why was the Moon chosen as the next destination in space? The practical reason was that it is Earth's nearest neighbor, but there was a sentimental reason, too—people had dreamed of going to the Moon for centuries.

Different Strategies

The Soviet Union and the United States used different strategies in the race to land the first humans on the Moon.

The Soviets sent a series of space probes to the Moon before attempting manned Moon missions. The Americans, on the other hand, tested the equipment that would be needed for manned missions to the Moon with manned test flights in Earth orbit.

Luna Program

The Soviet Union's Luna Program involved a series of unmanned space probes, which were sent to the Moon. In September 1959, *Luna 2* became the first man-made object to reach the Moon. *Luna 3* brought back the first images of the Moon's far side the following month. In April 1966, *Luna 10* became the first artificial object to orbit the Moon.

This is the Soviet space probe *Luna 2*, the first spacecraft to reach the Moon. It was destroyed in a crash landing.

Project Gemini

Unlike the Luna Program, the United States's Project Gemini was a series of manned missions in Earth orbit. The Gemini missions were designed to perfect the skills and equipment that would be needed on manned missions to the Moon, such as **docking** two spacecraft, testing spacesuits on spacewalks and spending up to two weeks in space.

The first docking in space took place on March 16, 1966 between *Gemini 8* and a spacecraft called the *Agena Docking Target*. This photo of *Agena* was taken from *Gemini 8* during its approach.

Mission Fact!

Gemini means twins in Latin and is the name of a well-known constellation (a group of stars which form a pattern). Project Gemini was given this name because its spacecraft carried two astronauts.

APOLLO: WALKING ON THE MOON

MISSION OBJECTIVE: To explore the Moon

NASA's Apollo Program was designed to land people on the Moon and bring them safely back to Earth. Six Apollo missions achieved this goal, and they are still the only missions ever to have done so.

Apollo 11

On July 20, 1969, American astronauts Neil Armstrong and Edwin "Buzz" Aldrin became the first human beings to set foot on another world. Armstrong described his "one small step" onto the surface of the Moon as "one giant leap for mankind."

About 600 million people (one-fifth of the world's population at the time) watched the event live on television.

▼ Apollo 11 astronaut Buzz Aldrin sets up an experiment to detect moonquakes. On the right is the lunar module named *Eagle*.

Mission Fact!

During the course of the Apollo Program, 12 American astronauts landed on the Moon. Altogether, they spent more than 80 hours exploring its surface. They also collected more than 838 pounds (380 kg) of Moon rocks.

Apollo 13

Just nine months after the first Moon landing, the world was gripped by another space drama. Two days after the launch of Apollo 13 on April 11, 1970, an electrical fault caused an explosion in the service module that crippled the entire spacecraft. Thousands of people worked around the clock to come up with a plan that would get the astronauts back to Earth safely. All three crew members landed in the Pacific Ocean on April 17 and were rescued by the USS *Iwo Jima*.

Apollo 13's command module, *Odyssey*, is hoisted onto the USS *Iwo Jima*, following the crew's rescue.

Apollo's Legacy

Based on studies of the rocks which the Apollo astronauts brought back from the Moon, scientists believe the Moon was created when a **protoplanet** collided with Earth when it was very young. Materials and technologies developed for Apollo are now used to make all kinds of things, including space blankets, food packaging, and cordless vacuum cleaners.

Timeline: Apollo Program

1967	1968	1969	1969	1970	1971	1971	1972	1972
A fire during a training exercise kills all three Apollo 1 astronauts in January.	In December, Apollo 8 becomes the first manned spacecraft to orbit the Moon.	Apollo 11 lands the first two men on the Moon in July.	In November that same year, Apollo 12 does the same.	The Apollo 13 crew is brought back home safely after an explosion cripples the spacecraft in April.	Apollo 14 lands on the Moon in January.	Apollo 15 lands on the Moon in July.	Apollo 16 lands on the Moon in April.	Apollo 17 lands on the Moon in December.

THE SPACE SHUTTLE PROGRAM

MISSION OBJECTIVE: To build a reusable spacecraft

Even before Apollo 11 landed on the Moon, NASA was planning its next project—a reusable spacecraft. Constructing such a spacecraft would make getting into space much cheaper. With this aim in mind, the Space Transportation System, or space shuttle, was born.

The Parts of the Shuttle

The winged **orbiter** is the main part of the shuttle. Two reusable solid rocket boosters provide most of the power for liftoff. The only part of the spacecraft that is not reusable is the giant external tank, which is released from the shuttle some time after lift-off and burns up as it falls through Earth's atmosphere. When a mission is over, the orbiter glides back to Earth. It can then be prepared for another mission.

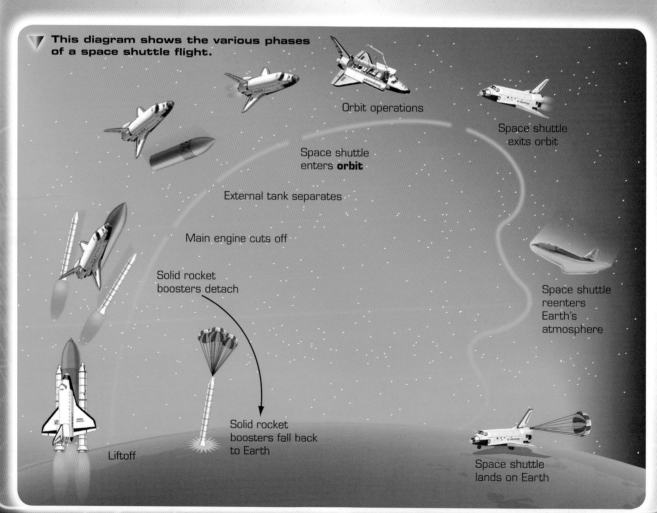

▼ **This diagram shows the various phases of a space shuttle flight.**

Orbit operations

Space shuttle enters **orbit**

Space shuttle exits orbit

External tank separates

Main engine cuts off

Solid rocket boosters detach

Space shuttle reenters Earth's atmosphere

Liftoff

Solid rocket boosters fall back to Earth

Space shuttle lands on Earth

▲ When a space shuttle needs to be moved from one location to the next, it hitches a ride on a modified Boeing 747, as *Discovery* did in 2005.

Mission Fact!

Six space shuttle orbiters were built: *Enterprise*, *Columbia*, *Challenger*, *Discovery*, *Atlantis*, and *Endeavour*. Of these, only *Enterprise* never flew in space. It was used as a test vehicle.

Shuttle Missions

The space shuttle can carry up to 63,500 pounds (28,803 kg) of **payload** into **low Earth orbit**. Since its launch in April 1981, the space shuttle's missions have included building, servicing, and transporting crews to and from the *International Space Station*, servicing the *Hubble Space Telescope*, and launching numerous satellites and space probes.

Shuttle Disasters

The space shuttle has flown more than 130 flights. Although most of these were successful, two of them ended in tragedy. In January 1986, *Challenger* exploded shortly after its launch. In February 2003, *Columbia* broke apart during **reentry**. Both disasters killed all seven crew members. Inquiries into the cause of each disaster stopped the shuttle program for a total of almost five years.

SPACE STATIONS

Space stations are designed to allow humans to live in space. Space station crews conduct scientific research that can only be performed in microgravity, such as studying the long-term effects of spaceflight on the human body.

Early Space Stations

The world's first space station was the Soviet Union's *Salyut 1*, launched in April 1971. Eight more Salyut stations followed. In 1973, the United States launched its first space station, *Skylab*. Both *Skylab* and the Salyut stations were constructed on the ground and had just a single module. The first station with multiple parts was the Soviet Union's *Mir*, which was put together in space between 1986 and 1996. *Mir* was decommissioned in 2001.

Mission Fact!

When *Skylab* **reentered** Earth's atmosphere in 1979, it broke up over the town of Esperance in Western Australia. The town fined NASA US$400 for littering. The fine was eventually paid on behalf of NASA by an American radio station, which raised the funds with the help of its morning show listeners in April 2009.

▼ **This photograph of the *International Space Station* was taken by a crew member onboard the space shuttle *Discovery*.**

▲ Canadian-born astronaut Gregory Chamitoff ponders his next move in a game of chess on board the *ISS*.

The International Space Station

Sixteen countries are working together to build the *International Space Station* (*ISS*), which is already the largest space station ever constructed. Assembly began in 1998 and is due to be completed in 2011. The *ISS* is made up of 15 separate modules. These include laboratories as well as docking ports, **airlocks**, and living quarters. When complete, the *ISS* will be 361 feet (110 m) long and will weigh almost 925,100 pounds (419,600 kg). It is so large that it can be seen from Earth with the naked eye.

Life Onboard the ISS

ISS crew members perform a variety of daily maintenance tasks to keep the station running smoothly. They spend the rest of their working day carrying out experiments. To counteract the effects of weightlessness, they spend up to two hours a day exercising. In their free time, they watch films, read books, play games, and even talk to their families on Earth.

THE HUBBLE SPACE TELESCOPE

MISSION OBJECTIVE: To get a better view of the universe

The problem with Earth-based telescopes is that Earth's atmosphere blurs visible light and partly or completely blocks other types of electromagnetic radiation. This problem can be avoided by placing a telescope in space.

The Hubble Story

The *Hubble Space Telescope* was named after American astronomer Edwin Hubble. Soon after its launch in 1990, scientists discovered a flaw in its main mirror, which took three years to repair. Once the telescope was working properly, it began sending amazingly clear and detailed images from space. The wait had been worth it.

Mission Fact!

The *Hubble Space Telescope* is the only space telescope designed to allow astronauts easy access to its scientific instruments for repair or replacement work.

How Does the Hubble Space Telescope Work?

Hubble is in low Earth orbit. Scientists on Earth can point the telescope at almost any target they want to study. *Hubble's* instruments are sensitive to ultraviolet, visible, and infrared light. Information from these instruments is beamed down to Earth via **satellite** and converted into pictures that astronomers can study.

Technicians assemble the *Hubble Space Telescope* at Lockheed Missile Space Company in 1985. *Hubble* is 43.5 ft (13.2 m) long, has a diameter of 14 ft (4.2 m), and weighs 24,500 lb (11,110 kg)!

This image taken by the *Hubble Space Telescope* shows two spiral galaxies interacting. The gravity of the larger one, on the left, is pulling the smaller one toward it.

Hubble's Discoveries

The *Hubble Space Telescope* has made an enormous contribution to the science of astronomy. It has given scientists a better understanding of the life cycles of stars and **galaxies**, and has shown that **black holes** not only exist, but that nearly all galaxies may have a black hole at their center.

Edwin Hubble
(1889–1953)

Edwin Hubble was an American astronomer. In the 1920s, he discovered that the universe contained not just one galaxy but billions of them, moving away from each other at incredibly high speeds. This discovery eventually led to the development of the **big bang theory**.

EARLY MISSIONS TO MARS

MISSION OBJECTIVE: To explore Mars

More spacecraft have been sent to Mars than to any other planet. This is partly because Mars is relatively close to Earth, and mainly because Mars is one of the places where we are most likely to find other life forms.

Is There Life on Mars?

Of all the planets in the solar system, Mars is the most like Earth. Its atmosphere is the most similar to Earth's. Both planets have polar ice caps, four seasons, storms, volcanoes, valleys, and canyons. Since they have so much in common, scientists have wondered for a long time whether life might exist on Mars too. Most missions to Mars have attempted to answer the question "Is there life on Mars now, or has it ever existed there?"

Mariner Program

The Mariner Program involved a series of robotic space probes that NASA sent to Mars, Venus, and Mercury between 1962 and 1975. *Mariner 4*, *Mariner 6*, and *Mariner 7* performed brief **flybys** of Mars. When *Mariner 9* orbited Mars in November 1971, it discovered craters, volcanoes, canyons, and evidence that water had once flowed on the planet's surface.

▼ **This meteorite from Mars, which was found in Antarctica in 1984, contains microscopic structures that some scientists believe might be fossilized bacteria.**

This image of Mars was taken by the *Viking 2* lander. The dish-shaped object is the antenna used to communicate with Earth.

Viking Mission to Mars

Launched in 1975, NASA's two Viking spacecraft each consisted of an orbiter and a **lander**. The mission objectives were to map Mars, analyze its atmosphere and surface, and search for signs of life. The Viking mission was a huge success, but it found no signs of life at either landing site.

Mission Fact!

When *Mariner 9* arrived at Mars, the surface of the planet was completely hidden by dust storms. It took a couple of months for the dust to settle and for the spacecraft to get clear images.

RECENT MISSIONS TO MARS

MISSION OBJECTIVE: To find out whether life ever developed on Mars

Early exploration of Mars showed that water once flowed on its surface. More recent Mars missions have followed the tracks left by this water, because where there is water there is more likely to be life.

Orbiters

NASA's *2001 Mars Odyssey* orbiter, launched in 2001, revealed large amounts of ice just beneath the surface in the northern polar regions of Mars. In 2005, NASA's *Mars Reconnaissance Orbiter* discovered that the planet's surface had once been very wet for a significant length of time. This means conditions may have been favorable for life.

▼ **This image compares the amount of ice in Mars's northern hemisphere during winter and summer. In winter, the ice is hidden by a thick layer of carbon dioxide frost, which is why the blue areas of ice appear smaller than in summer.**

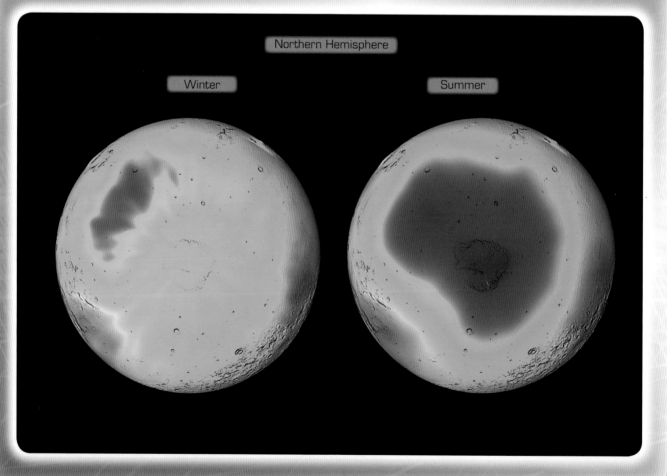

Northern Hemisphere

Winter

Summer

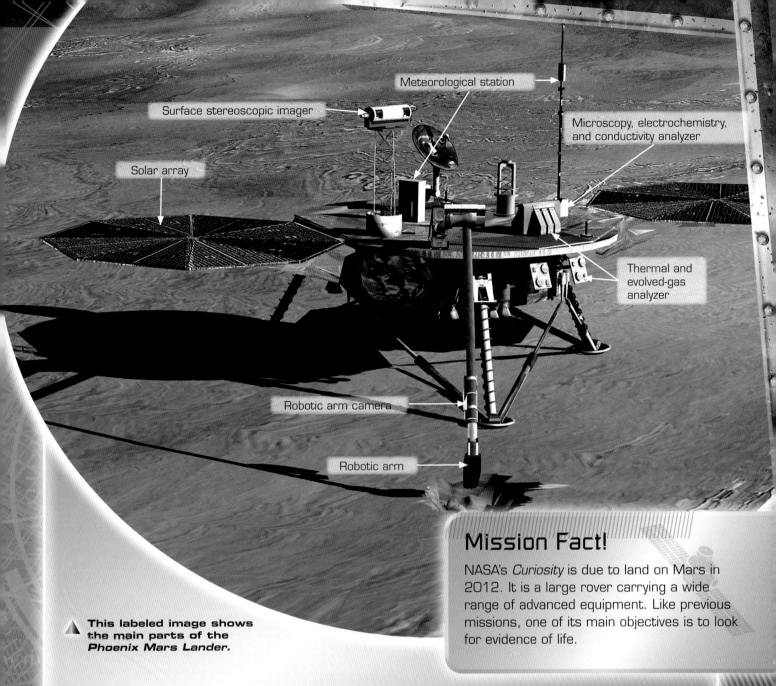

Meteorological station

Surface stereoscopic imager

Microscopy, electrochemistry, and conductivity analyzer

Solar array

Thermal and evolved-gas analyzer

Robotic arm camera

Robotic arm

▲ This labeled image shows the main parts of the Phoenix Mars Lander.

Mission Fact!

NASA's *Curiosity* is due to land on Mars in 2012. It is a large rover carrying a wide range of advanced equipment. Like previous missions, one of its main objectives is to look for evidence of life.

Landers and Rovers

In 2004, NASA's twin Mars Exploration Rovers landed on opposite sides of the planet. Three years later, NASA sent the *Phoenix Mars Lander* to Mars's northern polar regions in order to examine the ice discovered earlier by *2001 Mars Odyssey*. Both missions found more evidence that Mars may once have been capable of supporting life. *Phoenix* found salts that could act as nutrients for life forms and the rovers discovered evidence that ancient Mars had lakes and hot springs.

What Have We Learned from Missions to Mars?

Missions to Mars have shown that the planet was once relatively warm and wet, so it may have been capable of supporting life. From studying Mars, scientists have discovered that conditions on a planet can change dramatically. They are keen to learn even more about Mars because this will improve their understanding of Earth, its changing climate, and its ability to support life in the future.

MISSIONS TO VENUS AND MERCURY

Missions to Venus and Mercury face unique challenges. Venus is so hot and the pressure of its atmosphere is so great that spacecraft cannot survive there for long. As for Mercury, traveling there uses more fuel than a trip beyond the solar system, because a spacecraft's engine has to work against the pull of the Sun, or its gravity.

Venera Probes

Between 1961 and 1984, the Soviet Union sent a series of 16 probes called Venera to Venus. Despite the hostile conditions, 13 of the probes managed to send back useful data before malfunctioning, and 10 of them landed successfully on Venus. Later missions carried cameras that sent back photographs of the planet's surface. *Venera 4* was the first spacecraft to enter the atmosphere of another planet and *Venera 7* was the first spacecraft to land on one.

Pioneer Venus Project

The Pioneer Venus Project was an American mission. It consisted of two spacecraft, launched separately in 1978. One spacecraft was an orbiter, and the other carried four atmospheric probes. The orbiter mapped the planet's surface, using a radar imaging system, and studied its upper atmosphere. The probes were released from orbit, and as they fell toward the planet, they collected more detailed data about its atmosphere.

▼ **The *Venera 13* lander** captured this image of the surface of Venus before it malfunctioned due to extreme heat and pressure. The picture shows part of the spacecraft and a camera lens cap.

This artist's impression shows *MESSENGER* approaching Mercury. Its instruments are shielded by a sunshade.

MESSENGER

Launched in 2004, NASA's *MESSENGER* spacecraft has performed three **flybys** of Mercury, which revealed large amounts of water in the planet's upper atmosphere.

After entering orbit in 2011, *MESSENGER* will map Mercury and study its magnetic field, atmosphere, and internal structure. The instruments it will use to collect this information are listed in the table below.

Instrument	Purpose
Mercury Dual Imaging System (MDIS)	to capture three-dimensional (3-D) images of Mercury's surface features
Gamma Ray and Neutron Spectrometer (GRNS)	to map the distribution of different elements and discover whether there is ice at Mercury's poles
Magnetometer (MAG)	to study Mercury's magnetic field and search for magnetized rocks in the crust
Mercury Laser Altimeter (MLA)	to create 3-D maps of the planet
Mercury Atmospheric and Surface Composition Spectrometer (MASCS)	to identify gases in the atmosphere and minerals on the surface
Energetic Particle and Plasma Spectrometer (EPPS)	to study Mercury's **magnetosphere**
X-Ray Spectrometer (XRS)	to detect the materials that make up Mercury's crust
Radio Science (RS)	to find out how Mercury's mass is distributed by measuring slight changes in *MESSENGER's* orbit

MISSIONS TO THE OUTER PLANETS

MISSION OBJECTIVE: To explore Jupiter, Saturn, Uranus, and Neptune

Missions to the distant outer planets are truly epic journeys. The *Galileo* spacecraft spent six years traveling to Jupiter, the nearest of the outer planets, and it took Voyager 2 an incredible 12 years to reach Neptune!

Galileo's Encounter with Jupiter

NASA's *Galileo* was the first spacecraft to send a probe into Jupiter's atmosphere. The probe encountered high winds, clouds of ammonia (a toxic gas), and lightning up to 1,000 times more powerful than lightning on Earth. *Galileo* discovered what scientists believe to be a saltwater ocean under the frozen surface of one of Jupiter's moons, Europa, which could be capable of supporting life.

Mission Fact!

Galileo's fuel consumption was an average of 9.3 million mi (15 million km) per 0.26 gallon (1l). If a car were this efficient, it would use just four tablespoons of gasoline driving to the Moon and back!

▼ *Galileo* captured this image of a volcanic eruption on Io, one of Jupiter's moons, in February 2000.

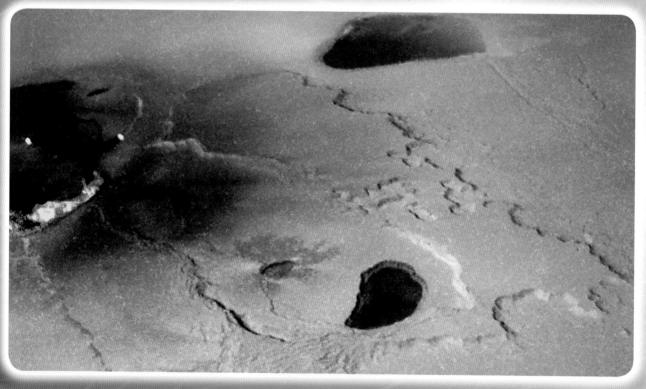

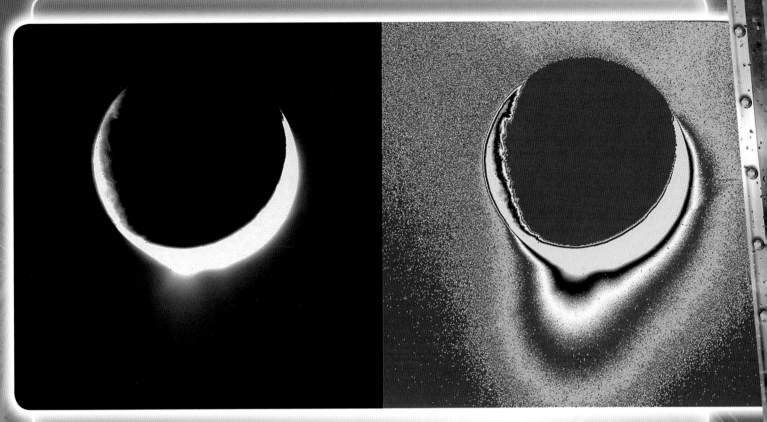

⚠ **Cassini–Huygens captured this image of Saturn's moon Enceladus, with plumes of ice and water vapor being blasted into space. A color-enhanced version is on the right.**

Cassini–Huygens: *Unlocking Saturn's Secrets*

The *Cassini–Huygens* mission to Saturn made many amazing discoveries, including a **hot spot** on the tiny moon Enceladus, where enormous plumes of ice and water vapor gush out into space. *Cassini–Huygens* also discovered that Saturn's largest moon, Titan, has rivers and lakes just like Earth. Instead of water, however, they contain liquid methane, a major component of natural gas.

Giovanni Cassini
(1625–1712)

Giovanni Cassini was an Italian astronomer. He discovered four of Saturn's moons and the gap in Saturn's rings, which was named after him (the Cassini Division). The *Cassini* orbiter is also named after him.

Voyager 2: *Flying past Uranus and Neptune*

Launched in 1977, *Voyager 2* flew past Uranus in 1986, finding 10 previously undiscovered moons. In 1989, it reached Neptune and revealed a series of fine rings that cannot be observed from Earth. Since leaving Neptune, *Voyager 2* has explored the outer reaches of the solar system and will continue its mission into **interstellar space**.

OTHER MISSIONS

MISSION OBJECTIVE: To explore other bodies in the solar system

The Moon and the planets are not the only places humans have sent spacecraft to explore. The missions on these pages were designed to study the Sun, a comet, and the asteroid belt.

SOHO

SOHO (*Solar and Heliospheric Observatory*) is a joint project of NASA and the European Space Agency (ESA). *SOHO* orbits the Sun in step with Earth at a distance of about 932,057 miles (1.5 million km) from Earth.

Since 1995, *SOHO* has been studying the Sun's internal structure, its outer atmosphere, and the **solar wind**.

▼ This image, taken by *SOHO*, shows a coronal mass ejection (CME). During a CME, more than a billion tons of matter shoot from the Sun at speeds of more than 621,371 mi (1 million km) per hour.

Mission Fact!

SOHO has discovered more than 1,500 comets, making it more successful than all other discoverers of comets throughout history put together.

Dawn

NASA's *Dawn* spacecraft is currently on its way to the main asteroid belt to study its two largest members, the **asteroid** Vesta and the **dwarf planet** Ceres. *Dawn* is due to reach Vesta in 2011 and Ceres in 2015. Both of these bodies are very old. By studying them, scientists hope to learn more about how our solar system was formed.

Deep Impact

NASA's *Deep Impact* spacecraft was launched in 2005 to probe beneath the surface of a comet. Like *Dawn*, its purpose was to discover more about the origins of the solar system, but by studying the make up of a comet. The main spacecraft carried a smaller impactor spacecraft, which was released into the path of Comet Tempel 1.

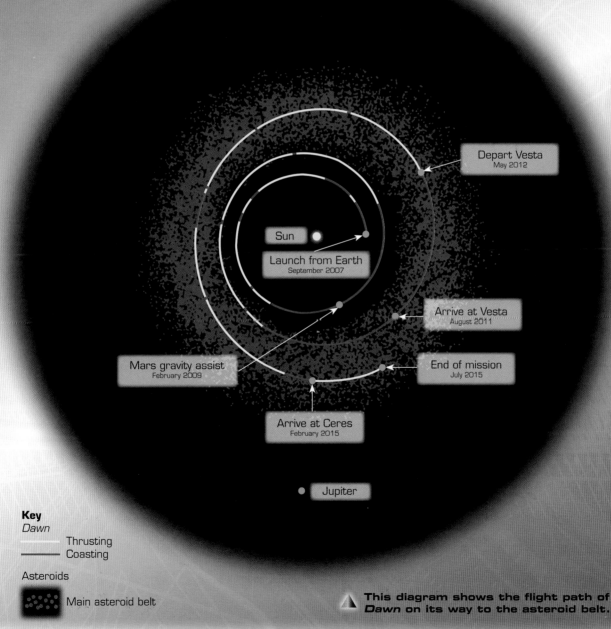

Depart Vesta
May 2012

Sun

Launch from Earth
September 2007

Arrive at Vesta
August 2011

Mars gravity assist
February 2009

End of mission
July 2015

Arrive at Ceres
February 2015

Jupiter

Key
Dawn
Thrusting
Coasting

Asteroids

Main asteroid belt

⚠ This diagram shows the flight path of *Dawn* on its way to the asteroid belt.

THE FUTURE OF SPACE MISSIONS

The Space Race ended in the mid-1970s with the joint Apollo–Soyuz project between the United States and the Soviet Union. Since then, many more countries have made the leap into space. In 2003, China became the first Asian nation to send astronauts into space. Recent developments in space exploration include the birth of space tourism and an increase in the number of privately funded spaceflights.

Where to Next?

In 2007, 14 space agencies from around the world released "The Global Exploration Strategy," a document that outlined their plans to work together to meet common goals in space exploration. These plans include establishing a permanent base on the Moon and sending the first humans to Mars.

Looking Farther Ahead

If the human population continues to expand, it will eventually need more space and more resources than are available on Earth. In the distant future, most space missions may be concerned with collecting resources and finding new places for humans to live.

▼ In this artist's impression of humans landing on Mars, the astronaut on the left is rappelling to take a closer look at the walls of a canyon. The astronaut on the right is setting up a weather station.

Mission Fact!

The first space tourist was American Dennis Tito in 2001. It was reported that he paid US$20 million to the Russian Federal Space Agency to join a Soyuz mission to the *International Space Station*. Tito spent more than a week in space.

GLOSSARY

airlocks
chambers in a spacecraft that allow astronauts to move in and out of the spacecraft without affecting its air pressure

asteroid
a rocky or metallic object that is between a few feet to more than 560 mi (900 km) across and orbits the Sun or another star

asteroid belt
a region between the orbits of Mars and Jupiter where most of the asteroids in the solar system can be found

atmosphere
the layer of gases surrounding a planet, moon, or star

big bang theory
the theory that the universe expanded from an extremely dense and hot state and continues to expand today

black holes
regions of space where gravity is so powerful that nothing can escape, not even light

comet
a collection of ice, dust, and small, rocky particles that orbits the Sun

docking
the process of joining one spacecraft to another

dwarf planet
a small planet-like body that is not a satellite of another body but still shares its orbital space with other bodies

electromagnetic radiation
waves of energy created by electric and magnetic fields

flybys
maneuvers in which a spacecraft passes close to a planet but does not enter its orbit or land

galaxies
large systems of stars, gas, and dust held together by gravity

hot spot
an area of active volcanism on the surface of a planet or moon

interstellar space
the region of space starting at the edge of the solar system and extending to the edge of the Milky Way galaxy

landers
spacecraft designed to land on a planet or moon

low Earth orbit
an orbit between 100 and 1,240 mi (160–2,000 km) above the surface of Earth

magnetosphere
the region around a planet or star that is affected by its magnetic field

microgravity
weightlessness, a phenomenon experienced when orbiting a planet

orbited
followed a curved path around a more massive object while held in place by gravity; the path taken by the orbiting object is its orbit

orbiter
a spacecraft designed to orbit a planet or natural satellite

payload
objects carried into space on a spacecraft, usually artificial satellites or space probes

protoplanet
a small planet in the early solar system that eventually collided with other protoplanets to form the present-day planets

reentry
the process of reentering Earth's atmosphere after a spaceflight

rovers
spacecraft designed to move around on the surface of a moon or planet

satellites
natural or artificial objects in orbit around another body

solar system
the Sun and everything in orbit around it, including the planets

solar wind
a stream of charged particles ejected from the upper atmosphere of the Sun

Soviet Union
a nation that existed from 1922 to 1991, made up of Russia and 14 neighboring states

Space Race
a period of competition between the United States and the Soviet Union in the area of space exploration

INDEX